A New Sy

Exercise for Infantry

Sir Richard Francis Burton

Alpha Editions

This edition published in 2022

ISBN: 9789356785335

Design and Setting By

Alpha Editions
www.alphaedis.com

Email - info@alphaedis.com

Contents

THE TARGET.

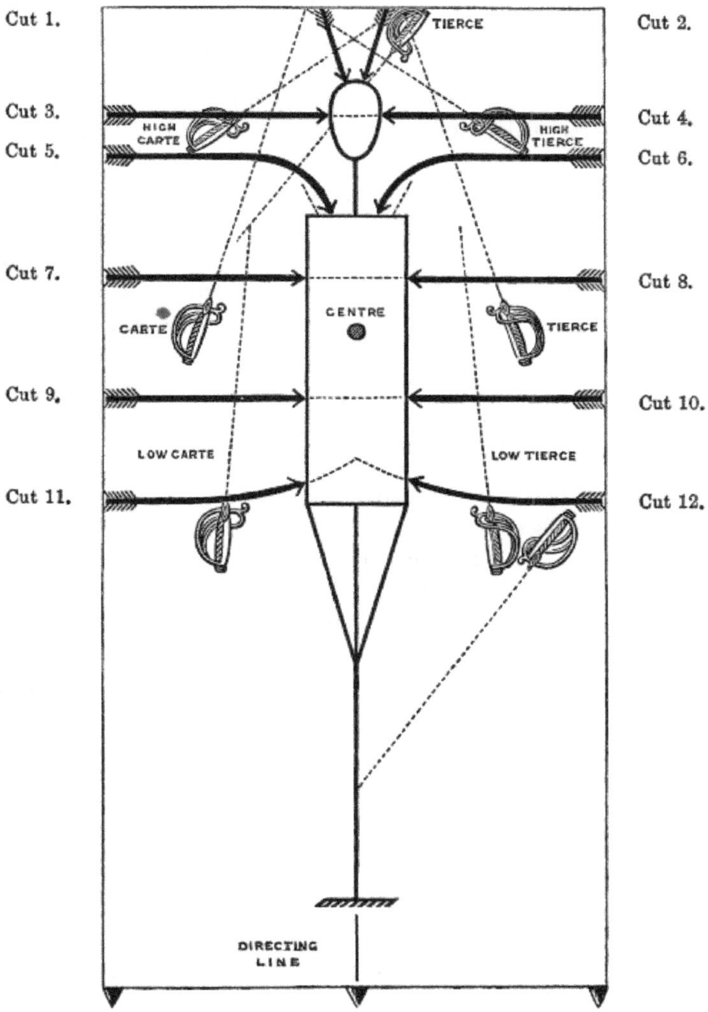

Tierce (or Outside Line).
Carte (or Inside Line).

INTRODUCTORY REMARKS.

BEFORE proceeding to develop my New Sword Exercise for Infantry, I would offer a few remarks upon the changes proposed in these pages. Whilst the last half century has witnessed an immense improvement in the projectile weapons of the civilized world, the theory and practice of the sabre or cutting arm have remained *in statu quo ante*; indeed, if there has been any change it is for the worse. The two systems authorized in the British army are completely behind their time. First and senior is the 'Infantry Sword Exercise' (with plates): Revised Edition, Adjutant-General's Office, Horse Guards. London: Printed under the superintendence of H.M. Stationery Office: 1874. The second is the 'Instructions for the Sword, &c. (without plates), for the use of Cavalry.' Adjutant-General's Office, Horse Guards. June, 1871.

The latter can be despatched very briefly. Despite the late date, it is as obsolete as the older system; it is, in fact, only the 'Infantry Exercise' with the addition of "pursuing practice," and "post practice"—the latter upon a sort of modern Quintain not made to revolve. So far, so good. The practised swordsman has little to learn when mounted, except the few modifications which he can teach himself. His real study is on foot. But some of the remarks appear not to have been written by a practical hand. For instance, we read (p. 27): "In delivering a forward thrust, very little force is necessary when the horse is in quick motion, as the extension of the arm, with a good direction of the point, will be fully sufficient." "Fully sufficient"—I should think so! The recruit must be carefully and sedulously taught when meeting the enemy, even at a trot or canter, to use no force whatever, otherwise his sword will bury itself to the hilt, and the swordsman will either be dragged from his horse, or will be compelled to drop his weapon—if he can. Upon this point I may quote my own 'System of Bayonet Exercise' (p. 27):—

"The instructor must spare no pains in preventing the soldier from using force, especially with the left or guiding arm, as too much exertion generally causes the thrust to miss. A trifling body-stab with the bayonet (I may add with the sword) is sufficient to disable a man; and many a promising young soldier has lost his life by burying his weapon so deep in the enemy's breast that it could not be withdrawn quickly enough to be used against a second assailant. To prevent this happening, the point must be delivered smartly, with but little exertion of force, more like a dart than a thrust, and instantly afterwards the bayonet must be as smartly withdrawn." In fact the thrust should consist of two movements executed as nearly simultaneously as possible; and it requires long habit, as the natural man, especially the Englishman, is apt to push home, and to dwell upon his slouching push.

The 'Infantry Sword Exercise' is nought but a snare and a delusion. Except in pagination, it is the same as the "Revised Edition" of 1845—the only difference or revision that I can detect is the omission of a short sentence in p. 26 of the older issue; it even retains the General Order of Lord Hill, 23rd April, 1842. Thus "Revision" is confined to the plates. In 1845 the figures wear the milk-pail shako widening at the top, the frock coat and the scales; the last edition, dated April, 1874, dons the tall modern chimney-pot, the tightly buttoned tunic with stiff collar and, like its predecessor, the sash and the scabbard. It is no wonder that the figures display an exceeding *gêne*, the stiffness of pokers, as the phrase is: here we might with profit borrow from the French or Italian artist.

I am opposed to almost every page of this unhappy *brochure*, especially to the "Seven Cuts and Guards" of the target; to the shape of the target—I never yet saw a man absolutely circular; to the grip of the sword; to the position in guard; to the Guards or Parades, especially the inside engaging guard (Carte); to the Lunge; to the angle of the feet, and to the system of "loose practice."

The "Cuts" will be noticed in a future page. Of the grip I may remark that the one essential, the position of the thumb, both in attacks and parries is, as a rule, neglected by the 'Sword Exercise.'1 As early as 1828, Müller made his *point d'appui* a grasp of the handle with the four fingers, the thumb being stretched along the back, in order to direct the edge, and to avoid the possibility of striking with the "flat." The only exception to this universal law is when doing the "Moulinet" movements, which will be explained farther on. Some professors, both with broadsword and small-sword, would stretch the index, when pointing, along the right of the handle. I have objected to this practice in the rapier and the foil: except when done to change position for relief, it serves merely to fatigue the wrist. But the proper use of the thumb, "*le pouce allongé sur le dos de la poignée*," which is troublesome at first, and which demands some study, especially from those who have acquired bad habits, is the base of all superior "counterpoint."2

The position on guard is a debated point. Many, indeed I may say most, of the moderns follow the rule of all the older swordsmen, namely, reposing two-thirds of the body-weight (as in p. 19 of the 'Exercise,' which, however, is an exaggeration) upon the left leg. The reasons usually given are that in this position the person is not so much exposed; moreover, that the centre of gravity being thrown back adds spring and impetus to the Lunge. We may remember how Cordelois (1862) made a step towards change in his fencing-schools at Paris. My objection to the old style is that the farther you are from your opponent, the longer and slower will be your attack; moreover, I have ever found, in personal practice, that it is easier and more convenient to "sit on guard" with the weight equally distributed on both haunches and legs. In fact, that the backward position is not natural any pair of thighs can ascertain for itself after trying it for five minutes: whilst the muscles of the right or forward limb are relaxed as much as possible, those of the left are tight strung, so as to do double work and threaten cramp. This single objection is serious enough to counterbalance any other claims to superiority.

First Guard. (Prime.)

("What to avoid.")

Again, there is no excuse for the guards in the 'Exercise.' The "Hanging guard" (p. 18, in the older issue p. 21) is the worst that can be imagined—a painful spectacle, a lesson of "what to avoid." The head ignobly cowers, and the eyes look up, in a forced and wearying position, when the former should be held upright, and the glance should be naturally fixed upon the opponent's eye and blade-point; the body is bent so as to lose our national advantage of height and strength, and the right fore-arm in such a position is, and ever must be, clean uncovered. Let the recruit, however strong may be his haunches, stand a few minutes in this "Hanging guard," and he will soon feel by his fatigue how strange, awkward, and strained it is. The Carte or inside Engaging Guard (pp. 19, 22), again, endangers the fore-arm. The Tierce or outside Engaging Guard (pp. 20, 23) holds the hand too low, and unduly shortens the arm, thus offering an undesirable amount of exposure; it is in fact not a Guard, but a bad parry in "low Tierce." Worse still is the Lunge (pp. 14, 17): here the body is placed bolt upright, instead of being easily bent, without exaggeration, to the fore, prolonging, as every man instinctively would do at his first attempt, the line of the left leg. The former position is not only fatiguing and "against the grain;" also shortens the reach and carefully places the opponent safely out of measure. Many swordsmen still contend for the stiffly upright position in Lunge:3 I am disposed to consider it a mere survival of the classical and artificial French school of arms, which aimed at opposing nature as sedulously as the Italian, who always leans to the fore, attempted to follow her dictates. Moreover, their arguments are founded upon the abuse, not the use, of the inclined pose which the body naturally assumes. In teaching the recruit it is well to see that he does not fall into the dangerous habit of throwing the chest forward (*poitriner*) to meet his opponent's point; but the truth of muscular motion must be consulted.

Finally, I would note the mistake of "loose practice" with the single-stick instead of the sabre; it probably arose from a mistaken economy in saving swords and paddings. Single-stick is a different weapon, a cane or light cudgel with a basket-hilt covering the back of the hand, like the imperfect guard of the Highland Clay-more; it is straight, not curved, and, as the rod has no edges, so in practice every blow equally represents a cut. Single-stick has merits of its own, but its practice is fatal to excellence with the broadsword, and even the 'Exercise' seems to recognize the fact, for the *guindés* figures are armed with officers' Regulation swords.

Inside Guard—(Carte).

(Weight all thrown back.)

Both 'Sword Exercises' carefully avoid naming "Tierce" and "Carte;" preferring "right" and "left" (of the Sword) or "outside" and "inside," as if such mysteries were too high or too deep for our national intelligence. I would again quote a few lines from my 'System of Bayonet Exercise' (Introductory Remarks, pp. 8, 9):—

"But why, it may be asked, should the English soldier be deterred by difficulties which every French voltigeur can master? We admire the intelligence of our neighbours in military matters: we remark that they are born soldiers, and that their men learn as much in four months as ours do in six. Is not this, however, partly our own fault? In my humble opinion we mistake the cause of their quickness, attributing to nature the effect of art. When our system of drill is thoroughly efficient; when the *Manual and Platoon* is much simplified, when a *salle d'armes* is established in every corps, and when the bayonet exercise becomes a recognized branch of instruction; then, I believe, we shall find our soldiers equal in intelligence to any others." These words were written in 1853; in 1875 I add, "When we enlist the right kind of recruit either by improving his condition and his prospects, not his pay, or better, far better, by securing a superior man through the conscription of modern Europe." We Britons are no longer physically divided from the total orb; nor can we afford to remain morally insulated and isolated. The logical effect of union with the outer world will be to make us do as the world does, and all our exceptional institutions, such as the system of volunteer recruiting, must sooner or later go by the board.

Outside Guard—(Tierce).

(A Parry not a Guard.)

Nor is the most modern French Treatise (pp. 229–256, *Manuel de Gymnastique et d'Escrime*, officially published by the *Ministre de la Marine et des Colonies*; Paris, Dumaine, 1875) "*Escrime au Sabre*" much superior to our home growth. The position of the left hand (pp. 232, 233) is bad throughout: it must slip during the Lunge and make the play loose. The retreat of the left leg (Fig. 5, p. 235) is carried to an extreme of caution. The body is always perpendicular in the Lunge, whereas the same volume shows (Fig. 16, p. 20) the trunk naturally inclining forwards. The Cuts are not double nor continuous, as they should be. The "Hanging Guards" (pp.

240, 244, 245) are deplorable. On the other hand, the *Manuel* (p. 231) places the thumb along, not around, the handle; the *moulinets*, the *enlevés*, and the *brisés* (presently to be explained) are good stuff, and, moreover, they are applied to the Cuts (p. 239). Finally, nothing can be better than the advice (p. 249), "Après avoir touché, retirer vivement le sabre en arrière en lui imprimant une direction oblique dans le sens du tranchant, de manière à *scier.*"

Of the points or thrusts with broadsword nothing will here be said: they belong to another order of things, and they should be studied in the fencing school.4 But the soldier must be taught that if his adversary attempt a thrust, the broadsword is easily disarmed. When the opponent comes to the position of pointing, that is, extends his blade, a sharp glissade along its length will make the grip fly out of his grasp. Another way of embarrassing the attack is to cut right and left at the hand, the wrist, or the fore-arm, when the adversary begins to present point.

General Lamoricière was a firm believer, as we all are, in the thrust, and the French Sword Exercise for Cavalry (p. 178 *Règlement Provisoire sur les Exercises de la Cavalerie*, officially published at the *Ministère de la Guerre*; Paris, Dumaine, 1873) justly remarks: "*Les coups de pointe doivent toujours être employés de préférence, comme exigeant moins de force et ayant un résultat plus prompt, plus certain et plus décisif.*" The reason of its confessed superiority to the Cut is as old as the axiom, "a straight line is the shortest way between two points." The Thrust describes a diameter, the Cut, a segment, of a circle and, with equal velocity, the Cut will traverse a distance occupying some two-thirds more of time than the Thrust. The French tactician therefore proposed to abolish the use of the edge for cavalry, thus traversing the instinct of the man-at-arms who, especially on horseback, loves to slash at his enemy, and who runs far less risk of entangling his blade. But he of course advocated a straight and tapering sword with no edge to speak of; indeed the cuirassier's *latte* is still a kind of rapier, but it is rendered useless by prodigious length and by the weight of the handle. The modern Italian School of Sabre uses, especially in single combat, all the *dégagements* of the *salle d'armes*: this is thoroughly illogical; the weapon is chosen because it is supposed to be less fatal than foil or rapier, and yet it is so used as to become even more deadly. I need hardly say that the weight and shape of the broadsword, together with the positions of guard, render pointing with it awkward in the extreme.5

I have now finished with the ungrateful task of criticizing, and I proceed to propose a system which it is hoped will be as severely criticized by others. It is only candid to state that its pretensions are high, that it contains two distinct novelties, the Manchette System and the Reverse or Back-cut; and, finally, that it aspires to be the first Treatise in which the broadsword is scientifically taken in hand.

SECTION I.
PREPARATORY INSTRUCTION WITHOUT THE SWORD.

§ 1. Preliminary.

NOTHING will here be said concerning the "goose step of the sword," the "Balance Motions," and the "Extension Motions," of the official 'Infantry Sword Exercise.' They are essentially a part of 'Squad and Setting-up Drill,' and as such they have been treated in several good manuals, especially by Serjeant-Major S. Bertram Brown: A 'Practical Guide to Squad and Setting-up Drill, in accordance with the Principles laid down in Part I., Field Exercise of the Army.' Adapted for the use of Recruits, Rifle Volunteers, Militia, Police Force, Schools, and Families: Illustrated with sixty-eight figures, representing each Stick and Club Exercise, Extension Motions, and Sword Exercise Positions. London: Allen and Co., 1871. 2nd Edition.6 Considered in a wider sense they belong to the Branch of Science so thoroughly developed in 'A Military System of Gymnastic Exercises for the Use of Instructors: Adjutant-General's Office, Horse Guards, 1862; Physical Education,' *Clarendon Press Series*, Oxford, 1869; and in 'Training in Theory and Practice' (London, Macmillan, 1874), by Archibald MacLaren,7 whose excellent code for the army, and whose influence with successive war ministers, as some one truly said, have aided largely in introducing that admirable training which is transforming the stiff, slow-moving grenadier of past times into the vigorous, rapid, and enduring soldier of the present day.

Squad drill is not likely to make a good swordsman, yet economy of time renders it a necessity. It must be practised first without, then with, weapons, after which those who show unusual capabilities should be taken individually in hand by the master. The latest French system (*Manuel, etc.*) divides the four lessons into two degrees: 1. Preparatory Movements; moulinets and simple attacks and parries. 2. Compound attacks and parries.

The formation of the squad is in the usual line, with open order at arm's length from the right or left. The men are then taught the three positions as follows:—

First Position. Second Position. Third Position.

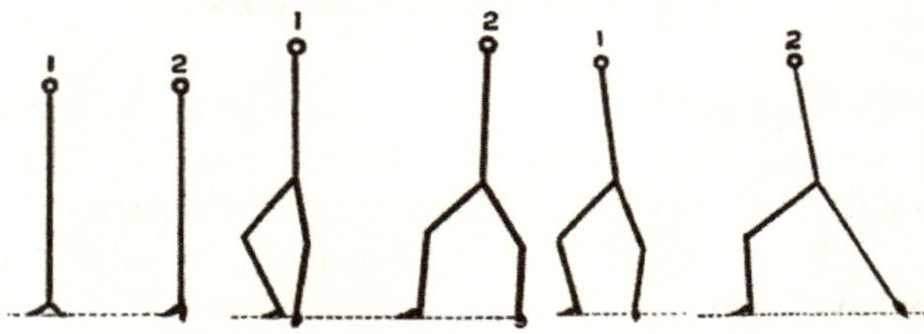

In two movements. In two movements. In two movements.

§ 2. First Position in Two Motions.

One.—Place the hands smartly behind the back, the left grasping the right arm just above the elbow, and the right similarly supporting the left elbow.

Two.—Make a *half-face right* by pivoting smartly on both heels, which must be kept close together; the feet at right angles; the left pointing to the front, the face looking towards the opponent, or the right-hand man, and the weight of the body balanced equally upon both haunches and legs.

Second Position in Two Motions (Guard).

One.—Bend the knees gradually till they are perpendicular to the instep, keeping the head and body erect, and both feet firm on the ground. The instructor must be careful that the knees do not incline inwards—a general fault.

Two.—Advance the right foot smartly about 20 inches in front of and in line with the right heel, and rest the whole weight of the body upon both haunches and legs.[8]

In the second position, that of Guard for the feet, care must be taken that the left foot remains firm on the ground, without shuffling or turning inwards or outwards. Many swordsmen find a better balance when the right heel is on a line with the hollow of the left foot.

Third Position in Two Motions (from Guard to Lunge).

One.—Advance the body slightly forward, and bring the right shoulder and knee perpendicular to the point of the right foot.

Two.—Advance the right foot smartly, about 20 inches, or double the distance of No. 2, Second Position (Guard), taking care that the foot does not overhang the instep; extend the left leg with a spring, the left foot remaining true and firm, and the left knee perfectly straight; let the shoulders expand and the body be profiled and slightly inclined forwards, or towards the opponent.

This is the position of the legs in the Lunge, and the greatest care must be taken to prevent the recruit learning it in a careless, shuffling way. Above all things he must accustom himself to separate the action into its two composing parts, otherwise the lower limbs will often take precedence of the upper (shoulder, arm, and hand), and the Lunge become worse than useless. When recovering guard the contrary is the case; the left knee must be bent before the right foot is moved, and the latter should exert a slight pressure on the ground; at the same time the body must be drawn backwards, not jerked upwards.

These measures of Guard (20 inches) and of Lunge (40 inches) are best fitted for average-sized men; in exceptional cases they must be shortened or lengthened according to the stature and stride of the recruit. The rule for guard is the measure of two foot-lengths; the Lunge doubles that span; and the least vigorous men require the greatest distances.

These movements must be learned, first in slow, and afterwards in quick and in double-quick, time; the same may be said of all practice with and without the sword. *Squad attention!* and *Stand at Ease!* need hardly be explained. The recruits' muscles soon become fatigued by the unusual and monotonous exercise, causing them to remain too long in one position; the easiest way to relieve them is to change front, making the left leg stand on guard and lunge, as a left-handed fencer would do. This double practice is as useful and recommendable in fencing and broadsword play as in bayonet exercise: it gives additional balance to the body, it equalizes the muscular strength of both sides, and it makes the soldier feel that if his right arm be disabled he can still depend upon his left.

The word *Steady* must not be used as a command: it should be a caution given at the completion of any part of a practice with the view of correcting faults.

§ 3. Attacking, Advancing, and Retiring.

Single Attack.—Raise the right foot well off the ground and beat smartly with the whole sole, the greatest force being upon the ball of the foot, and the least upon the heel.

Double Attack.—The same movement made twice. The instructor should carefully avoid the directions of the 'Infantry Sword Exercise,'—*first with the heel, then with the flat of the foot.* Nothing jars the leg more than this use of the heel; it is a bad habit to use it for anything but "pivoting."

Advance.—Smartly advance the right foot about six inches and bring up the left as nearly as possible to the same distance. The soles must just clear the ground, and the toes be kept on a straight line with the knee, and never turned inside or outside. Neglect of the latter precaution leads to a loose, unsteady, and slovenly style which, easily learnt, is hardly to be unlearnt.

Single Attack.—As before.

Retire.—Move the left foot lightly to the rear about six inches, and let the right foot follow it. Recruits are uncommonly apt to "step short," and this can be remedied only by making them retire for considerable distances. The weight and balance of the body must be equally distributed on both haunches and legs, not resting upon the left, which can serve only to give cramp.

Double Attack.—As before.

Front.—Resume the position of "Attention."

SECTION II. PREPARATORY INSTRUCTION WITH THE SWORD.

§ 1. Explanation and Use of the Target.

THE Target prefixed to these pages explains itself. The shape is oblong, the frame measuring 6 feet by 3, and the figure 5 feet 8 inches by 1 foot. As the latter represents the opponent, the centre should be about 4 feet from the ground, the height of the recruit's breast. Perpendicular to the foot of the figure in each Target a horizontal line is drawn, forming for the feet, the legs, the body, and the arms, the "directing line" of the scientific schools. At a distance of 10 feet the recruit is placed in the position of "Attention," with his left heel on the line, so that at the command "First Position" his right foot may cover it.

The parallelogram shows the direction and the numbering of the Cuts, concerning which further details will presently be given. They should be regulated according to the lines described upon the Target; nor should the recruit be practised in any other mode until he has gained the proper direction of the blade.

Nothing need be added to the directions of the 'Infantry Sword Exercise' (pp. 12, 13, 14), as regards the movements subject to the following words of command: much, on the other hand, with great advantage, might be taken away, and the result would be the increased efficiency that results from simplicity.

Draw Swords (should be much abridged; after the modern French School, pp. 165, 166: *Règlement Provisoire*, &c.);

Slope Swords;

Return Swords (should be simplified);

Stand at Ease;

Attention;

Prepare for Sword Exercise;

Right prove Distance;

Slope Swords;

Front prove Distance; and

Slope Swords.

At the order, *Stand on Guard*, the recruit having assumed the Second Position, No. 2, falls on Guard: the pommel of the sword fronts his right breast; the point is directed at his opponent's right eye; his right arm is extended with an easy bend at the elbow; the wrist is inclined, with the knuckles slightly turned upwards, to his own right, so as to cover him in case of a straight thrust, and the left hand is placed upon the left flank just below the ribs, with the fingers to the front and the thumb to the rear.

The several guards (parries) are learned by holding the sword opposite to and in the inclination of the dotted lines which have sword-hilts attached to them; the recruit is thus taught from the Target the angle of the blade and the position of the wrist.

The Target directs the recruit *how* to make the Cuts and to form the Guards, but not exactly *where*; this must depend upon how the opponent acts during the attack and the defence. Cuts 1, 3, 5, 7, 9, and 11 (odd numbers) are all from Carte, which the 'Infantry Sword Exercise' calls *Inside*. The corresponding even numbers (2, 4, 6, 8, 10, and 12) are from Tierce, or *Outside*. The same nomenclature applies to the Guards or parries.

When the recruit thoroughly understands the use of the Target he need no longer be practised in front of it; but the instructor (sword in hand) should consider it a sure guide and reference for correctly forming the Guards and for giving a suitable direction to the edge when making the Cuts.

§ 2. The Moulinet.9

This rotation movement should be learnt before the recruit proceeds to the Cut.

There is nothing better for "breaking," as the French say, the recruit's wrist than this sweep of the sword; and the style of a swordsman may always be known by his Moulinet. We will divide it into three kinds, viz. (1) horizontal, (2) diagonal, and (3) vertical; the latter again may be either (a) ascending or (b) descending; but as the second (diagonal) is a mere modification of the first and the third, it will be sufficient to notice only two; these are:—

1st. The horizontal movement, or *Moulinet* proper, circling the sword round the head. The grip is held as lightly as possible, chiefly with the thumb and the first finger, resting the pommel upon the palm, and carrying the nails upwards. The blade should be moved as horizontally as it can be, with the back just clearing the swordsman's crown: it should describe, not a true circle, but an oval with a long diameter in the directing line to the centre of the Target through the heels or ankles of the recruit. Finally, the point should be lanced or thrown out, as it were, towards the opponent's face. Evidently it may be done in twoj ways, first, from right to left, which I will call the "Tierce Moulinet" (*Moulinet à gauche*); this is by far the easiest and the more habitual, corresponding with Tierce "Counter," opposition, or describing with the blade a circle round the adversary's blade, in the fencing school. The reverse movement ("Carte Moulinet," *Moulinet à droite*), from left to right, requires, like the Counter of Carte, much more practice.

In these directions "right and left" apply to the right and left of the swordsman's wrist.

2nd. In France the term "Moulinet" is mostly applied to these two rotations of the sword round the head, but we will extend it to all circlings of the point. The vertical form is also made from the hand in Tierce (Outside Guard), the blade is brought sharply round with the back towards the breast and left shoulder, and returns to its original position; we will call this the "Inside Moulinet," having reference to the performer, not the adversary. The "Outside Moulinet" is when from

"Tierce or Outside Guard" the blade passes along the right shoulder, it is simply the former done in the outer line.

Horizontal Moulinet.

Again the "Inside Moulinet," which ends with the Cut from above downwards (the French *enlevé*), may be inverted so as to cut from downwards upwards (the *brisé*). The same may be done with the "Outside Moulinet," when the wrist must be turned upwards, and the Cut given in the ascending line. This difficult movement should be practised in order to ensure a flexible wrist, but it exposes the whole arm. In the four latter "Cuts," the one invariable rule is to circle the point as vertically as possible. The French *Manuel* (pp. 234, 235) gives: 1, the *enlevé* cutting from above downwards; and it may be either *à gauche* (Tierce Moulinet) or *à droite* (Carte Moulinet); 2, the Moulinet proper; and 3, the *brisé*, cutting from downwards upwards, thus reversing the *enlevé*; and this also may be done *à gauche* (Tierce Moulinet) or *à droite* (Carte Moulinet).

Vertical Moulinet.

The "Moulinet" should be practised first without, then with, the sword, and on foot, before attempting it on horseback. In the earlier stage the recruit must turn the hand, with the arm nearly extended, in the horizontal and vertical movements, without stiffness and displacement of the elbow. In the second he may, if no Target be procurable, work before a cross chalked on the wall so as to secure horizontality and verticality. Finally, the soldier will combine the two, Tierce and Carte, by passing rapidly from one to the other.

Whilst practising the Moulinet the recruit must be taught the two main divisions of the sword-blade. Fencers have introduced an immense complication into this simple matter; and some have proposed eight parts: for broadsword it is sufficient to divide the length. The "Feeble," or weak half, is that contained between the point and the centre; this, the proper part for the Cut or attack is ground to a thinner edge, and consequently is more liable to an injury from another sword if the Cut be not very true. The "Fort," or strong half, is from the centre to the hilt, and upon this we must rely for defence.

A few hours' practice and a few pressings upon the different parts of the blade under the surveillance of the instructor will teach the recruit the high importance of this lesson. He will learn that in opposing the adversary's sword the strength of the defence decreases from the hilt upwards in proportion as the Cut is received towards the point; and that, *vice versâ*, it increases from the point downwards to the hand. The strongest man cannot "force in" the opponent's Guard if the Cut or Thrust be received upon the part near the handle. With a true Guard the ordinary fencing foil can turn off the thrust of a musket and bayonet weighing 10 lbs. The practised swordsman always attempts, when attacking, to gain with his "Fort" the "Feeble" of the opponent's weapon, in which case the superior leverage will often beat down the parry; and this manœuvre should be carefully practised by men of superior muscular strength. The Cuts must, as a rule, be delivered within eight inches of the point and at the "centre of percussion,"[10] so that the sword may clear itself and the arm escape a "jar."

The two virtues of the Cut are its trueness and its velocity. Unless true it will become a blow with the flat that would shiver to pieces any brittle Eastern blade. Assuming the *vis viva* or force of a moving body to be its weight multiplied by the square of the velocity, let us suppose a strong man cutting with a sword weighing 4 lbs., to which he can give a velocity which we will call 1, or 4 × 1 = 4: a weaker man who applies double the velocity to a 2 lb. sword will thus produce a momentum of 8, doubling the force of the blow. But let the stronger man take the lighter sword, evidently he will obtain a higher velocity, which we will assume at 3: in this case the effect will be 18. Thus the power of the Cut is enormously increased by increased velocity, but much less by increased weight in the moving body.

§ 3. The Cuts.

The Target prefixed to the 'Infantry Sword Exercise' gives Seven Cuts, an insufficient number. The German systems add an eighth blow perpendicularly upwards, when the whole of the swordsman's arm from wrist to shoulder would be completely at the opponent's mercy.

The French *Manuel* has only seven, viz. the *Coup de Tête*; 2, the *Coup de Banderole*; 3, the *Coup de Figure à droite*; 4, the *Coup de Figure à gauche*; 5, the *Coup de Flanc*; 6, the *Coup de Ventre*; and 7, the *Coup de Manchette.*

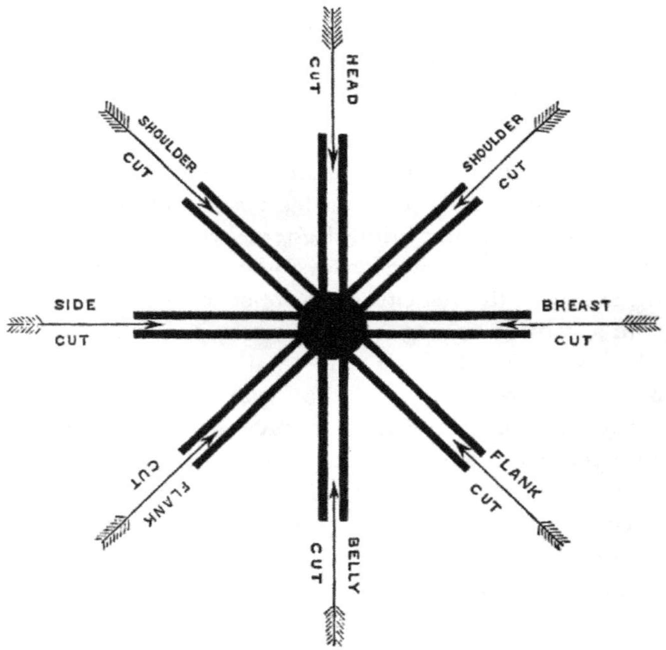

German System.

The subjoined diagram shows the Twelve Cuts11 which serve to "loosen" the rigid arm of the recruit.

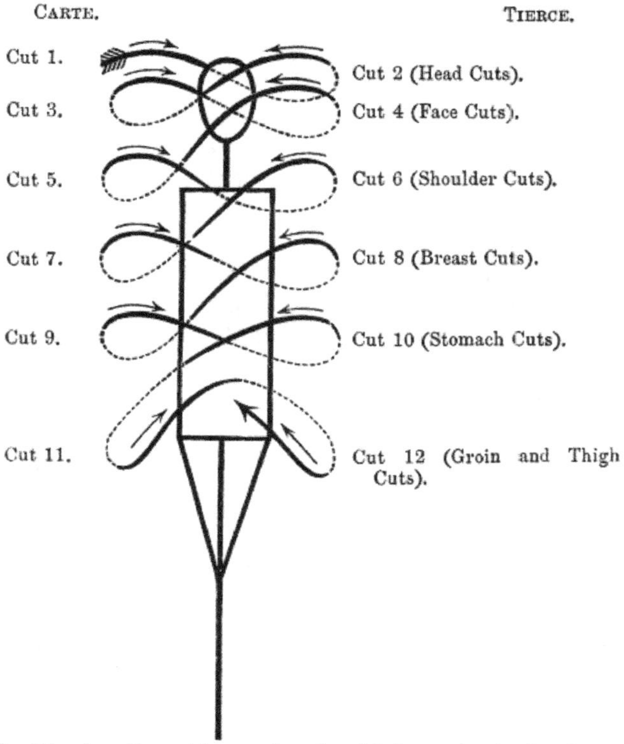

The Twelve Cuts (shown by the thick strokes), the dotted lines denoting the course of the blade in "Moulinet", or rather in "Semi-Moulinet."

CARTE.	TIERCE.
Cut 1.	Cut 2 (Head Cuts).
Cut 3.	Cut 4 (Face Cuts).
Cut 5.	Cut 6 (Shoulder Cuts).
Cut 7.	Cut 8 (Breast Cuts).
Cut 9.	Cut 10 (Stomach Cuts).
Cut 11.	Cut 12 (Groin and Thigh Cuts).

The figure represents the opponent; the thick lines show the direction of the edge when cutting; and the dotted continuations denote the course of the blade when describing the several "Moulinets."

The Cuts should be continuous, the regular succession always beginning from Carte or the Inside, that is, from the rear of the left shoulder. As in the "Moulinet," the less the arm is bent and the sword-hand is moved from the line of direction (to the front), the greater is the value of the movement. The recruit, who must walk before he runs, should deliver the whole dozen in continuous sweep without pause, but at first very slowly, till, by the proper and timely use of the wrist, the Cuts lead into one another. The more advanced swordsman, whose pliability of strength is free from contractions and other vicious habits, should practise the series of twelve with increased rapidity till the blade whistles through the air. All the Cuts should be given strong, with the edge leading well forwards and with the arm extended to its utmost in the delivery.

The following are the Twelve Cuts:—

I. and II. These Cuts are made, after falling into Tierce or Outside Guard, from above downwards at the opponent's head. In No. I. the point, beginning as usual from the left shoulder (Carte), describes a full circle ("Inside Moulinet," the *brisé à gauche* of the French *Manuel*), the hand moving as little as possible so as to cover the body; the knuckles turned up and the blade passing close to the breast: it finishes by delivering a vertical Cut, with the "Feeble" close to the point, at the right half of the adversary's crown. No. II., which follows without interruption, reverses the process; the knuckles are turned down and the blade sweeps past the right shoulder (*brisé à droite*); ending with the left half of the opponent's head. The latter Cut is by far the more difficult to make without moving the hand, but it is good practice for "breaking" the wrist.

III. and IV. The horizontal face-cuts, also beginning from the left (Carte), an invariable rule, and ending with the right, that is, at the adversary's left cheek. The reason of this practice is to make the movement habitual to the recruit; cutting from left to right always causes less exposure of the inner wrist than cutting from right to left.

V. and VI. The slanting shoulder-cuts, also from above downwards (Nos. 1 and 2, or rather 2 and 1, of the 'Infantry Sword Exercise,' pp. 14, 17, and the *Coups de Banderole* of the *Manuel*); describing two diagonal Moulinets, first from left to right, and then from right to left. The sword again makes a double "Moulinet" with the edge downwards, and descends first upon the opponent's right and then upon his left shoulder.

VII. and VIII. The horizontal breast-cuts, parallel with the face-cuts, and, like them, delivered with the blade as horizontal as possible.

IX. and X. The horizontal stomach-cuts, parallel with, and lower than, the breast-cuts.

XI. and XII. The slanting groin or thigh-cuts, diagonally from downwards upwards; in fact, the reverse of the shoulder-cuts (Nos. 4 and 3 of the 'Exercise,' and the *brisés* of the *Manuel*). In these diagonal Moulinets, the elbow must not be bent; the hand should deviate as little as possible from the directing line under pain of dangerous exposure; and the two movements should follow each other without a break.

Whenever the recruit fails to carry the edge well forward in making the attack, he should be practised slowly and repeatedly in combining the opposites, as Head-cut (No. 1) and Thigh-cut (No. 12), Head-cut (No. 2) and Thigh-cut (No. 11), and so forth. The instructor must see that the edge leads on to the respective lines of the Target, the point being darted out at the end of each cut.

The Cuts will be practised first in No. 2 "Second Position" (Guard), and afterwards in No. 2 "Third Position" (Lunge).

Prime, or Hanging Guard.

§ 4. The Engaging Guards, or Engagements.

As the 'Infantry Sword Exercise' has a deficiency of Cuts, so it has a superfluity of "Engaging Guards." I have already expressed my opinion concerning the Guard (p. 18 of 1874) popularly called the "Hanging Guard." Even with the best position, the head erect and the eyes looking straight and not upwards, it is utterly faulty; it displaces the arm and the sword, and, as no serious attack can be made directly from it, it necessitates a movement entailing a considerable amount of exposure. It is now chiefly confined to students' duels with the German Schläger, wherein slitting the opposing nose, which can be done with a mere jerk upwards, is the swordsman's highest aim and ambition.

The "Engaging Guards" are thus reduced to the two following:—

Tierce (or outside) Guard; defending the outer lines, arm, shoulder, back, and flank. The recruit having assumed the "Second Position" (No. 2), brings the pommel of his sword to the centre of his right breast; opposes the point to the adversary's right eye; extends his right arm with an easy bend of the elbow; inclines the wrist with the knuckles upwards to his own right, so as to cover himself in case of a straight thrust, and places his left hand upon his left flank with the fingers to the front and the thumb to the rear. In Tierce of course the edge of the sword is to the right or outside.

Engaging Guard in Tierce (Outside).

Carte (or Inside) Guard. This movement defends the inner lines, chest and stomach; the knuckles are turned down; the opposition is made to the left, and the edge is carried in the same direction.

Engaging Guard in Carte (Inside).

When engaging in guard (joining weapons), the swords should meet each other about eight inches from the points. If the distance is diminished the opponents are "out of measure" (or distance); if increased, they are "within measure." The recruit must be taught slightly to press upon the opponent's blade, but not to rest upon it; by this "opposition" his hand and wrist will be more ready to follow the weapon during the attack. Thus also the "Engaging Guards," Tierce, and Carte (outside and inside) afford protection preparatory to the movements for offence and defence. The eye must be fixed upon the eye and the hand or the blade-point of the opponent, not upon the eye only.

Guard may be partly defensive when the bust is advanced and the point approaches the opponent, or it may be purely protective when its sole object is the "parry."

The right-handed recruit must be taught always to attempt Engaging in Tierce,12 with his opponent's blade in the outer line (*sur les armes*). The reason is simply that in the reverse position (*dans les armes*), the fore-arm, from the elbow to the wrist, is comparatively unguarded; whereas Tierce facilitates the defence of the "low lines" (i. e. those below the wrist). Tierce therefore has invariably the advantage with the sabre, as Carte carries off the palm with the small-sword, the foil, and the rapier.13 But the right-handed man engaging in Tierce puts his left-handed opponent in Carte; and the latter, if a skilful sworder, will manœuvre, by withdrawing his blade, by coupés or degagements over the point, and by other feints, to regain the ground of vantage. The best treatment of this case is to make a time-cut in Seconde ("inner Moulinet," or *brisé à gauche*) at the adversary's knuckles, a movement which will presently be explained.

§ 5. The Guards or Parries.[14]

The 'Infantry Sword Exercise' proposes Seven Guards, a number which can hardly be reduced for practice on the drill-ground or in the schools: the *Manuel* contains the same number, including one for the Point. But of the seven no less than five are "Hanging Guards," and Nos. 3 and 4 serve only to defend the inside and the outside of the advanced leg. This limb requires no assistance of the kind: an able swordsman never exposes his head and shoulders by cutting so low, and, if he does, the leg can be smartly withdrawn (*parade retrograde*, or *en échappant*), rendering the attack not only useless but dangerous to the assailant. Even in fencing, "low thrusts," that is, at the body below the wrist, are never made, for fear of the "Time" being taken, until the upper line has been closed by a feint. In our Single-stick practice the first thought seems to be to attack the advanced leg—which may be well enough for Single-stick.

Lunge and Cut in Carte (Inside).

The following are the full number of guards or parries in which the edge must invariably be used: they are evidently dividable into two; (1) Head (with face) Guards, and (2) Body Guards:—

I. Prime (p. 38), so called because it is the "first" position of defence after drawing the blade, that which the unpractised man would naturally assume to defend his head. It is the 7th Guard of the 'Infantry Sword Exercise.' In practice the point is more inclined to the horizontal line than when the blade is unsheathed; the edge is carried somewhat inside or to the left; the arm is shortened and so raised that the eyes look under it, but the head remains upright. The recruit must be careful *not* to "bend the body;" *not* to "draw in the chest and neck;" and *not* to "bring the left shoulder a little forward." The defect of Prime is its being a "Hanging guard," rendering the riposte or reply difficult, and modern practice prefers "High Tierce."

II. Seconde (4th Guard), so termed because following Prime: the arm is extended, the edge is carried to the outside or to the right; in practice the hilt is lowered, and the point, threatening the opponent's loins, is depressed to the half of a right angle. This position must be learned for the sake of feinting: as a parade it is not much used, because it defends only the hip and leg, and a good swordsman will never expose himself to exceeding danger by making low cuts. Modern practice prefers "low Tierce."

III. Tierce (2nd Guard) has been described (p. 39) under "Standing on Guard" and "Engaging Guards;" it defends the outer lines, arm, shoulder, and back.

IV. High Tierce is a head-guard: the hand is raised to above the shoulder to the maximum level of the swordsman's right eye, and the blade is carried at an angle of 45° with the edge up and the point to the left.

V. Low Tierce is a flank-guard; the arm is shortened, the hand is depressed six inches; the opposition is to the outside, and the point is held vertically or almost vertically, as the attack demands.

VI. Carte (1st Guard) has been described (p. 40) under "Engaging Guards," as defending the inner lines, chest and stomach. For the purposes of parrying, the arm is withdrawn till the elbow, almost touching the belt, forms an equilateral triangle with the hilt and the left side.

VII. High Carte is a head-guard like high Tierce: the hand is raised to the left of the left eye, and the blade, crossing the face at an angle of 45°, carries the edge up, and the point to the right.

VIII. Low Carte is a stomach guard. As in Low Tierce the arm is shortened, the hand is depressed six inches; the opposition is to the inside, and the point is held vertically or almost vertically, as the attack demands.

In practice the advanced swordsman will confine himself to Tierce and Carte with their natural modifications. He will consult his own feelings about the head-guard, abolishing Prime in favour of High Tierce or High Carte, and he will prefer Low Tierce or withdrawing the leg (*rassemblement*) to using Seconde. Of these movements the simplest are always the best. When parrying, the sword-arm must invariably be drawn for defence nearer the body, and the grip should be sensibly tightened to receive the cut. No strength is necessary when making the parries: I cannot accept the "Sforzi" or guard-forcings of the neo-Italian broadsword school, dry blows upon the blade, which, intended to disarm, are essentially dangerous.

The Guards or Parries will be practised like the Cuts, first in the "Second Position" (Guard), and afterwards in the "Third Position" (Lunge).

SECTION III. THE MANCHETTE OR FORE-ARM PLAY.

§ 1. Preliminary.

THE recruit is now sufficiently advanced to begin the system of Manchette, which, as it is the most valuable part of sword-drill, has been practised the least, and should be practised the most. A swordsman thoroughly trained in this section does not allow the opponent to deliver a cut. It is certain that hand and wrist, short-arm and elbow, are capable of as many different attacks and defences as the whole body: these are the parts most prominent, most exposed, and consequently most readily made the *point de mire*. Yet this true and simple secret of the broadsword has been universally neglected, or rather not worked out: in England we content ourselves with the parades technically called *retrogrades*, that is, withdrawing the limb from the assault, by shortening the arm and, sometimes, by retiring the right foot either near to, or up to, or behind the left heel: even this evasion which cannot expect to pass for a Guard, is not described nor figured in the official 'Infantry Sword Exercise.'15 In France, and even in Italy where most subjects are exhaustively treated, Manchette is dismissed with a few careless words. The *Manuel* gives to the *Coup de Manchette* only these few lines: "*Exécuter un enlevé* (vertical Moulinet from above downwards) *en arrière à droite, et arrêter le sabre vis-à-vis le milieu du Corps, le tranchant en dessus, le poule légèrement à droite; diriger l'enlevé de manière à empêcher en arrêtant l'avant bras, l'exécution d'un coup de tête.*" Capitano Settimo del Frate (p. 50, *Istruzione sul Maneggio e Scherma della Sciabola*) in one of the latest works on swordsmanship contents himself with the following desultory observations:

"Manchett" (*sic*) "can attack the fore-arm either above or below, according as the opponent gives an opening.

"Manchett is generally used against an adversary whose guard is defective. By merely extending the arm with a turn of the wrist, this attack may readily succeed should the opponent neglect to provide against it.

"One of the most dangerous guards against Manchett is Tierce; the surest is High Seconde, which indeed is also the best parry adapted to this system of attack."

The first member of the last paragraph is sensible; the second is thoroughly fallacious. As has been stated, the right-handed man must always engage in Tierce, and, as will presently appear, Tierce is the safest, indeed the only safe guard against Manchette cuts. Another Italian writer of our day describes and figures the "Position of the weaponed arm to escape the arm-cut" (*Colpo di braccio*), with the elbow-joint left clean open. The 'Infantry Sword Exercise' limits itself (p. 30) to these few lines: "If opposed to the Small Sword (*sic*, meaning straight sword or rapier) have recourse to Cuts *Three* (No. 13 of this system) and *Four* (No. 11), directing them at the arm, by which means there is every probability of the cuts taking effect, as it must always come within range of the edge, before the point can be sufficiently advanced to reach your body: if the above cuts are quickly given and continued, they will also be found advantageous in advancing against the Small Sword, as they constitute an attack and form a defence at the same moment; but should the opponent be the most skilful and quickest (*sic*) in his movements, then it is best to retire whilst giving them, cautiously preserving the proper distance, so that each cut may just reach the fore part of his arm." The French content themselves with single oppositions of Tierce and Carte. But why multiply instances of ignorance?—they would fill many a useless page.

Finally I meditated upon the comparative humanity of "Manchette," of disabling the opponent by an arm-cut, rather than laying open his flank or his head. During single rencontres in the field, especially at the end of Indian battles, it is so often necessary to put *hors de combat* some unfortunate, whose pluck or sense of honour induces him to prolong the hopeless attack.

These considerations led me to reflect seriously for a number of years upon the Jeu de Manchette, the Colpi all' avambraccio, or fore-arm play, which has been so much neglected by master-swordsmen. At last an unlooked-for opportunity, a short study in the *Salle d'armes* of Herr

Balthasar Reich of Trieste, enabled me to reduce it to a system, and present it to the public.

I should premise, however, that the following observations are intended for professional men. It is therefore necessary only to name and number the Direct Cuts, the Guards and the Feints, the Reverse Cuts, and the Time Cuts of Manchette, as in most cases the simplest mention will suffice. The proficient will at once perceive that I offer a mere outline of the system whose many details must be learned by long practice. It is enough to give first principles: the minutiæ could not even be noticed without stretching description to a wearisome length.

There is no objection, I have said, to teaching squads of recruits all the simpler preparatory matter: the Three Positions; the Moulinet; the Engaging Guard, and the Guards or Parries. At a certain stage of progress, however, especially when beginning Manchette, the quick and intelligent soldier, who is likely to qualify himself as a master, must be instructed singly.

§ 2. The Direct Cuts in Manchette.

The following are the direct attacks in Manchette, simple and compound; all are done from the "Engaging Guard," the Lunge being here inadmissible.

I. *Carte de Manchette.*—Extend the sword-arm to the full length and deliver the cut, with a flip as it were, at the opponent's fore-arm, between the elbow and the wrist. This can be done with the back of the blade (Reverse Cut) under circumstances presently to be described. No. I. is useful if the adversary unwisely engages you in Carte otherwise (from Tierce) it must be avoided, as he easily parries by withdrawing the arm and replies with a Tierce Cut.

II. *Carte de Manchette and Cut Tierce.*—This movement is No. I. followed by a close rotation of the point ("Tierce Moulinet"); if, however, the circle be too small, it will not clear the sword-guard.

III. *Double Carte de Manchette and Cut Carte.*—No. III. is to be done when the opponent, as he generally will after an attack of No. II., successively parries Carte and Tierce. It is simply the double of No. I., and thus the "Tierce Moulinet" cuts, of course, inside the arm.

IV. *Double Carte de Manchette and Cut Tierce.*—Useful when the adversary parries Carte, Tierce and Carte; it is the double of No. II. and thus cuts outside the arm.

No. II. guards the arm and is therefore unexceptionable. Nos. III. and IV. are dangerous, because, like No. I., when opposed to an agile hand, they may lay the wrist open to a Time Cut.

The two first and all four against a slow unready swordsman may be varied by combinations with *coupés*, or passing the blade sharply over the adversary's point. For instance, if the adversary come too wildly to the Tierce parade of your double Carte and Tierce (No. III.), *a coupé* will reach his arm in Carte.

A golden rule which cannot be repeated too often is that all the Manchette-Cuts in Tierce (outside), either from above or from below, must be as nearly vertical as possible, whilst all the Cuts in Carte (inside) should be as horizontal as they can be made. The reason is simply that these positions cover the arm and render the attack less dangerous.

§ 3. The Guards (Parries) and Feints in Manchette.

The Guards of the Target will be found sufficient for parrying all attacks in Manchette. The soldier, however, should especially practise the retrograde parades, that is withdrawing the right fore-arm with and without the right leg.

Feinting with the broadsword is necessarily more simple than with the foil, being generally confined to *Coupés* and *Secondes*. The neo-Italian school of sabre uses, I have said, the fencing movements, but it is at best a bastard style. If the opponent attempts to "degage," that is to pass his point under your blade from Tierce to Carte, or *vice versâ*, retire by withdrawing the right heel to the left, and cut at the arm which his movement has exposed.

The *Coupé*, the reverse of the degagement, passes the point over, not under, the opposing blade; this legitimate feint, used in every school, may be effected in four several ways.

I. *One*. From the usual engagement in Tierce pass the blade over the opposite point, just clearing it, and cut inside. The two movements raising and dropping the point should be as rapid as possible.

II. *One, Two*, a double *Coupé*, with the cut in Tierce.

III. *One, Two, Three*: as with the foil; against a nervous opponent the cut should be made at the face with a dart and a jerk (the Italian *Slancio*); against a slow player the cut may be Carte de Manchette.

IV. *One, Two, Three, Four*; like the former, but cutting in Tierce: to be attempted only with the most unready of opponents.

The two latter may be combined with a breast (inside) or shoulder (outside) "Moulinet" between the penultimate and the last (cut) movement; but these long feints are radically vicious, because they lay the swordsman open to Time Cuts. They are, however, useful, as will appear in making the Reverse Cuts.

Perhaps the Seconde-feints are better than the *Coupés*.

I. *One*: the simple Seconde Cut.—Make a little more opposition in Tierce, sweep the blade past and along the breast; (inside Moulinet, or the *brisé à gauche*) and, lowering the hand a little, cut upwards with a jerk and a flip. The nearer the swordsman's own body his blade circles the better, because the cut will be more in the vertical line: if it be much out of the perpendicular the opponent can "take a time" in Carte. The Moulinet serves also to embarrass the adversary and to add strength to the cut. This simple and most valuable movement must not be confounded with the old-fashioned Seconde cut at the leg: the latter is objected to, as I have said, by swordsmen; the parry is too easy, and the ripost far too dangerous.

II. *Feint Seconde.*—From Tierce make a short and sharp movement to Seconde with the knuckles turned upwards; the opponent will probably come to the Seconde-parry, thereby exposing the fore-arm. You then cut Tierce perpendicularly as usual, from above downwards (the *enlevé*), either without or with a breast "Moulinet."

III. *Feint Seconde, Feint Tierce and Cut Carte*, with two short, sharp movements, and deliver the horizontal cut in Carte.

IV. *Feint Seconde, feint Tierce and Cut Seconde*, from downwards upwards, always with a breast "Moulinet."

At times the two first feinting movements in Nos. III. and IV. may be done more emphatically: this of course makes the movement slower, but it is a variety which embarrasses an adversary accustomed only to short, quick action.

§ 4. The Reverse or Back Cuts in Manchette.

As the Manchette system has been strangely neglected, so the Reverse or Back Cut may be pronounced unknown to the majority of the profession: the latter, instead of utilizing the "false edge" of the blade, still lose time and incur great danger by turning hand and wrist in using the true edge, especially when "Cutting within the Sword."[16] More extraordinary still, although almost all the civilized world prefers what is technically called the "flat-backed and spear-pointed" sabre, yet no one seems to think of employing, or even of sharpening one of the most important parts of the weapon.

The Regulation blade with the false edge, that is to say, the blade sharpened from the point to the Centre of Percussion, about one-third of the length, was introduced into England about 1844, and the first specimens were made by the late Henry Wilkinson, acting with the late Henry Angelo, then Superintendent of Sword Exercise.[17] This back-edge of the blade should be ground to the sharpness of a razor. When practising the Reverse Cuts (*Revers* or *Rovescio*), the handle is held loosely with the thumb and the two first fingers, and the wrist and fore-arm should bring the blade up with a jerk, the grip being at the same time sharply tightened. Practice will soon enable the swordsman to deliver a strong "drawing" cut, equal to the Thrust-cut of the so-called "Damascus" blades. This valuable movement has the immense merit of not uncovering the swordsman, and what makes the sabre so rude a weapon is that every movement of attack, in the old systems, lays the body open by raising hand and point when a blow is to be given. With the Reverse Cut no such dangerous process is necessary; the point is still directed at the opponent whilst the cut is being delivered. Finally, it is always unexpected by the opponent who has not practised it, and although it rarely begins an assault, except against thin experienced, nor should it be done alone as a rule, it may either follow or conclude every attack, feint or "time."

I. *The Half-Feint* (*Revers de dessous*, *Rovescio di sotto*, or *Revers von unten*) is done thus.—When in Tierce extend the arm as if intending to cut Tierce; the opponent makes an opposition of Tierce; drop the point, and cut sharply upwards with the false edge at his fingers, wrist, or fore-arm, drawing the blade towards you and keeping the point opposite the adversary's breast. This movement is one of the neatest known, and it is sure to succeed with one who does not expect it. The first part of the feint, or dropping the point, may lead to a cut with the true edge, but this movement, which is still practised in the schools, involves delay by turning the hand. Again, it may be combined with the inside (breast) or outside (shoulder) Moulinet.

II. *Feint Seconde and Cut upwards.*—This movement may be varied by feinting Tierce and cutting upwards.

III. *One-two-three.*—This is not the succession of simple *Coupés*, the dangerous movement before described. No. 1 *Coupé* shifts the hand from Tierce to Carte with the nails up; No. 2 turns the nails down, still remaining in Carte; and No. 3 delivers the Reverse Cut, of course in Carte, where it is least expected.

When the point is passed well under and within the sword-arm it is very difficult to parry the horizontal Reverse Cut in Carte. The true edge may be used, but again it wastes time by turning the hand.

IV. *The Pass*, properly called "*en passant.*"—From Tierce make a feint-movement in Seconde, and, when the adversary attempts to parry it by lowering the point, turn the knuckles up (in old Tierce), sweep the blade over his sword-arm and as close as possible to your right leg from left to right with the arm well raised, and, returning from right to left with a similar sweep, but with the blade held higher, cut, in Carte, with the false edge and close to the point, inside his wrist. Unskilfully attempted, this feint is equally dangerous to both, but it will do yeoman's service in the hands of a practised swordsman. The true edge may be used, but that involves a change of position and the delay of turning the hand with the knuckles downwards. Some make a double sweep, and, after the second movement, cut outside or in Tierce—the exposure is too great, unless confronted by an unusually phlegmatic temperament.

§ 5. The Time Cuts in Manchette.

The Time Cut is the flower of the Manchette system, as the Manchette is of the broadsword; and it is, perhaps, the part least capable of being taught in books. When well mastered it never allows the opponent to raise his arm without imminent risk, and, even if it fail, the intention, once recognized, tends greatly to cramp and embarrass the adversary's play. The natural man cuts as if he were using a stick or a club, and the preliminary movement lays open the whole of his body; indeed, exposure, I have said, is the main danger of every attack with the sabre, however closely and skilfully conducted. A cut through the muscles of the fore-arm, either inside or outside, causes the sword instantly to be relaxed and dropped; the man in fact is hamstrung in the upper works.

I. *Carte de Manchette.*—When the opponent from Tierce makes a *Coupé* or any attack in Carte, stop further movement by a Carte de Manchette, a horizontal Cut in Carte. The same may be done with the false edge, in which case the blade should be advanced as far as is possible; and this is to be preferred because it loses less time.

II. *Parade Retrograde and Cut Tierce.*—When the opponent from Tierce attempts a Manchette in Carte withdraw the arm (*parade retrograde*) and deliver the vertical Cut in Tierce downwards at his extended arm; both movements being combined in one. It is not necessary even with the tallest man to withdraw the right leg; the Cut will amply suffice. This Tierce Cut serves to defend from all attacks when the Guard does not cover the adversary; and it has lopped off many a careless arm. If slowly done it becomes a mere parade and ripost.

III. *The Reverse Cut-upwards*, *Revers en montant*, *Rovescio montante*, Ger. *Revers montant.*—You feint in Seconde; the opponent comes to its parry and replies in Tierce; you withdraw the arm, leaving the heels as they were, and cut upwards with the false edge, tightening the grasp of thumb and fore-fingers as much as possible. This movement is especially useful; it is one of the best of Time Cuts, when the adversary indulges in long and complicated feints and false attacks. It may be done with the true edge, but the latter is less safe.

IV. *The Time Pass*; which is merely "The Pass" turned into a Time Cut. When the opponent attempts a "Manchette" or any movement in Seconde, and expects you to reply by a time Cut in Tierce with the true edge, turn the knuckles up (in old Tierce), sweep the blade over his sword-arm as close as possible to your right leg, from left to right, with the arm well raised, and returning from right to left with a similar sweep, but with the blade held higher, cut in Carte with the false edge and close to the point inside his wrist. The true edge may be used, but, again, it wastes time. The double sweep possible as in "The Pass," but it causes too much exposure.

This Time Pass may also be done with the hand held high in Prime or rather "demi-circle" with the nails turned up, the arm outstretched, and the point lowered. In this case the leg must be shifted till the fore heel touches the rear heel, so as to give additional height to the hand. This is not a Reverse or Back Cut as you use the true edge; it is in fact one of the old movements called "Cutting within the Sword."

§ 6. Résumé.

The following is a synoptical table of Manchette or Fore-arm play, showing the Cuts, the Guards (Parries) for the Cuts, and the Riposts or replies that should follow each Parade. The Instructor will remember that instead of Prime we use High Tierce or High Carte, and for Seconde Low Tierce or withdrawing the leg.

Cuts.

	Y.	ST.
1. Carte de Manchette.	IV. (Carte).	II. (Seconde).
2. Ditto and cut Tierce.	IV. and III. (Tierce).	III.
3. Double Carte de Manchette and cut Carte.	IV., III. and IV.	II.
4. Double Carte de Manchette and cut Tierce.	Parade Retrograde by withdrawing arm.	III. or IV.

Cuts.

1. Half-feint.	II. or III.	III. or IV.
2. Feint Seconde and cut upwards.	II.	Cut with false edge upwards.
3. Feint Tierce and cut upwards.	III. and II.	II.
4. One-two-three, and cut upwards.	Parade Retrograde.	III. or IV.
5. The Pass.	II. and I. (Prime).	III.

Cuts.

1. On all Cuts in Carte.	Parry with time in IV. (Carte de Manchette).	IV.
2. On feints in Carte ending with Cuts in	Parade Retrograde.	III. or IV.

Tierce.		
3. On Cuts in Tierce.	Reverse Cut upwards.	III.
4. On Reverse Cut upwards.	II. and III.	IV.
5. On Cuts in Seconde.	The Time Pass.	III.
of *Coupé* in Manchette.		
1. Single *Coupé.*	III. or IV.	II.
2. One-two („).	IV. and III.	III.
3. One-two-three.	II., III. and II.	III. or IV.
4. One-two-three-four.	Parade Retrograde.	III.
of *Seconde* in Manchette.		
1. Simple Seconde.	II.	III.
2. Feint Seconde and cut Tierce	II. and III.	III. or IV.
3. Feint Seconde, feint Tierce, and cut Carte.	II., III. and II.	III. or IV.
4. Feint Seconde, feint Tierce, and cut Seconde.	Parade Retrograde.	III. or IV.

CONCLUSION.

I will end this system of Manchette with the words of old Achille Marozzo, written some three centuries and a half ago: "I would that ye swear upon your sword-hilts never to use this knowledge against me, your master." But, in lieu of insisting that my readers never teach it without obtaining formal permission, I only hope that they will favour me by spreading it far and wide.

APPENDIX.

IN p. 26 allusion has been made to an improved form of sabre handle; it was first attempted by the Capitano Settimo del Frate in the work before alluded to. The gallant officer's Plates show that in the Italian cavalry-sword the upper portion of the handle is at least horizontal, whereas in ours it droops backwards and downwards, giving the grip additional facility for slipping out of the swordsman's grasp. The author's remarks[18] being even more applicable to the English military sabre; I give them at full length.

"The equilibrium of the sabre, and the facility of firmly grasping the handle, are the two prime requisites for a good weapon.

"When properly balanced and easily held, the sword calls for less exertion of strength; and the quickness and true direction of the Cuts are greatly facilitated. In direct proportion to the economy of force, we find the swordsman enabled to continue his exertion.

"However well made and scientifically poised be the blade, it is subject to several variations of equilibrium according to the position in which it is held.

"The nearer the centre of gravity approaches the hilt, the lighter and the better balanced will be the weapon, and *vice versâ*.[19] Therefore:

"*It should be our principal object to effect this improvement without changing the proper centre of percussion and the other requisites for offence and defence.*"

The following Plates fully explain the author's meaning.

Fig. 1.

Sabre handle actually used by Italian cavalry.

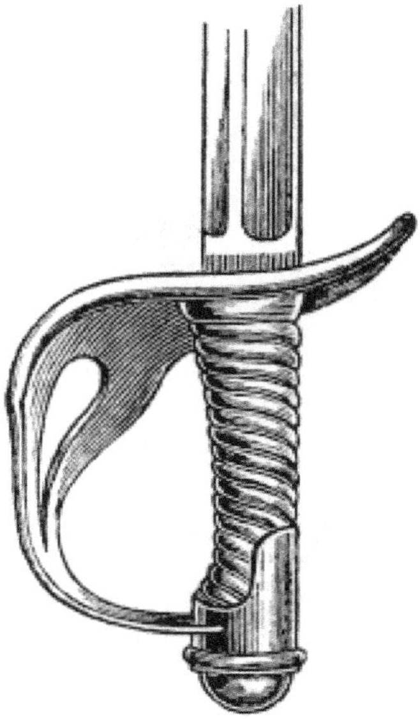

Fig. 2.

Capt. Del Frate's improvement.

Fig. 3.

Capt. Del Frate's last modification.

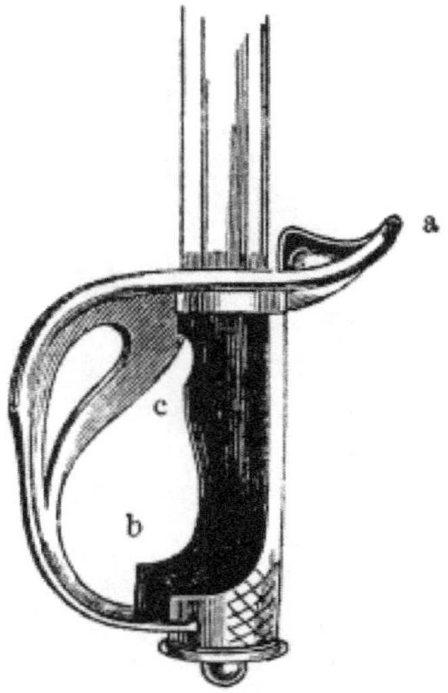

a. Thumb-plate.
b. Rest for the little finger.
c. Support for the index finger.

Fig. 4.

Improved handle with thumb-guard.
(R. F. Burton.)

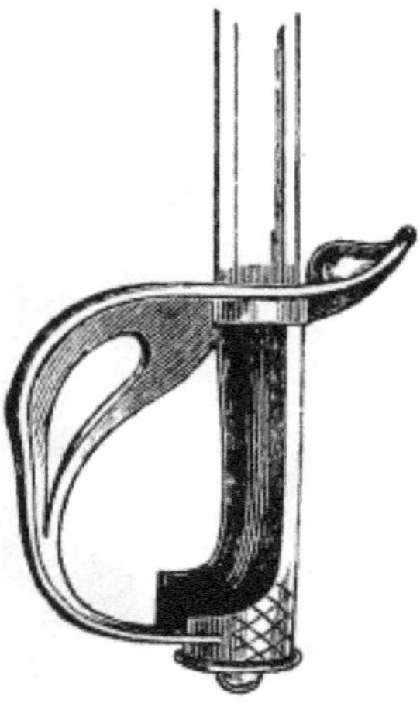

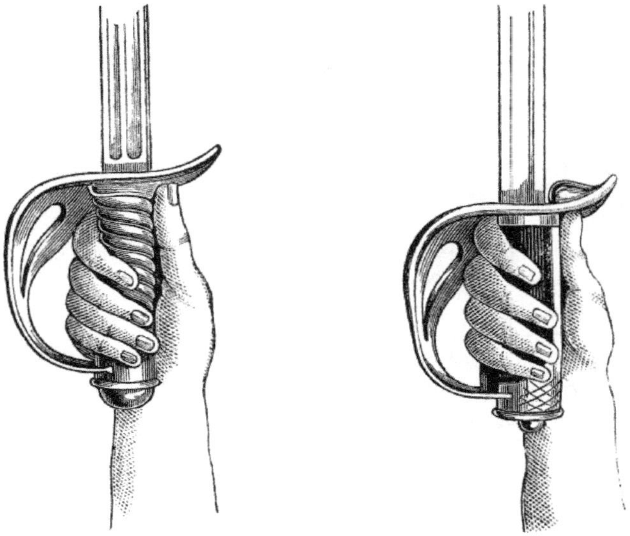

The hand grasping the actual handle. The hand grasping the modified handle.

I would further modify his Fig. 1, so as to give more fulcrum to the hand. The thumb-plate should be made weighty and the guard light, otherwise the blade will be over-balanced, that is, heavier on one side than on the other. It need hardly be said that the grip before going into battle should be whipped round with thin whipcord, or better still, with web-cloth.

FOOTNOTES

1 The exceptions are in "*Right Prove Distance*" (p. 13) and No. *Seven* Cut (p. 16). In the other Cuts the thumb "grasps the handle."

2 The French divide *l'Escrime* into two parts: (1) *Escrime à l'épée*, or *Escrime pointe*; and (2) *Escrime au sabre*, or *Escrime contrepointe.*

3 The question is considered at great length in my forthcoming volume entitled 'The Sword:' here it is sufficient simply to state results.

4 When every regiment shall have its *salle d'armes*, the fencer will modify his own fencing thrusts to suit the clumsier weapon. I do not, however, see any reason why the three Points of the Infantry Sword Exercise should not be delivered in the *posizione media* of the Italian school, with the thumb upwards and extended along the back of the sword-handle: nor why, as in the French *Manuel*, they should not be reduced to a single *Coup de Pointe* (p. 239), which is thus described. "*Baisser la pointe du sabre à hauteur de la poitrine et déployer le bras en tournant la main, le pouce en dessous, le tranchant du sabre en dessus.*"

5 As Mr. John Latham justly says ("The Shape of Sword-blades," 'Journal of the Royal United Service Institution,' vol. vi.):—"The proper shape for a thrusting sword is pre-eminently straight." The Clay-more, for instance, moving in a direct line, cuts a hole exactly the size of the blade; the Regulation sword, slightly curved, widens it to about double, and the bent scimitar and the Talwár, to five or six times, thus meeting with five or six times the resistance to its penetration. Mr. Latham is again quoted in another part of this System.

6 My only objections to this volume are the two following:—

(*a*) The author *will* "throw the whole weight of the body on the left leg." (Fig. 2, p. 69.) Yet in his Introductory Remarks

(p. 5) he sensibly says, "To the haunches, as to the common centre of motion of the human figure, are ultimately referred all the movements performed in military tactics" (and swordsmanship); "as just poise is important to the correct exertion of action, whatever it may be, it is necessary that poise or balance be studied, understood, and tried in all positions. It is clear that bodily action cannot possess compass, power, and ease, unless the movement be made justly and correctly upon the haunches, as on a central pivot. If the movement have not compass, power, and ease; force and endurance will not be found in the Military act."

(*b*) In the Lunge our author not only keeps the body "perfectly erect," he even inclines it backwards whilst he allows both feet to abandon the perpendicular in the most slovenly way: see Fig. 2, p. 70, and Figs. 1 and 2, p. 71. The same is the case with the official 'Infantry Sword Exercise.'

7 My old friend and instructor set out upon a thoroughly scientific principle, and the able way in which he has worked out his system will entitle him to the gratitude of the *posteri.* Having established the fact that in all our popular athletic, as opposed to gymnastic, exercises, our walking and running, cricket and football, fives, tennis, and racquets, and especially rowing—which has advanced as an art but has declined as an exercise—we circumscribe the line of muscular operation by giving the greatest share of the work to the lower limbs, and by developing one half to the injury of the other; he resolved to cultivate the whole by a wider and more varied range of training; hence he supplemented "Recreative exercise" by "Educational exercise," and hence his systematized national gymnasia, which, taken up by H.R.H. the Duke of Cambridge and by the late Sidney Herbert, have been introduced into the military stations of the Cardwell system, into Oxford and Cambridge, and into all our public schools, with one "base exception"—Eton.

Mr. MacLaren, in his 'System of Fencing.' &c. (p. 9), sensibly advocates "resting the weight of the body equally upon both legs." He also lowers the right hand in the Lunge (p. 11), and (*ibid.*) he throws the trunk forward, perhaps with a little exaggeration.

8 The 'Infantry Sword Exercise (see the figures over the Target representing the "Preparatory Positions"), "Second

Position in 2 Motions," makes No. 2 turn the left knee out instead of carrying it square to the front; the same may be remarked in "Balance Motions" (No. 4).

9 The Moulinet (Ital. Molinetto) is even on horseback a favourite movement with French sabrers (See *Règlement Provisoire*, &c., Tome I. Titres I. et II.). It is divided into—

1. "*À gauche Moulinet*" (1 temps, 2 mouvements). The directions are: "À la dernière partie du commandement, qui est MOULINET, étendre le bras droit en avant de toute sa longueur, le poignet en tierce et à hauteur des yeux."

"Baisser la lame en arrière du coude gauche pour décrire un circle d'arrière en avant ... et se remettre en garde."

2. "*À droite Moulinet*" (1 temps, 2 mouvements). "À la dernière partie du commandement, qui est MOULINET, étendre le bras droit en avant de toute sa longueur, le poignet en quarte et à hauteur des yeux."

"Baisser la lame en arrière du coude droit pour décrire un circle d'arrière en avant ... et se remettre en garde."

3. "*À gauche et à droite Moulinet*" (1 temps, 2 mouvements). "À la dernière partie du commandement, qui est MOULINET, exécuter le premier mouvement de *à gauche Moulinet.*"

"Exécuter alternativement et sans s'arrêter sur aucun mouvement le Moulinet à gauche et le Moulinet à droite."

"À gauche et à droite = MOULINET."

4. "*À droite et à gauche Moulinet* (1 temps, 2 mouvements). À la dernière partie du commandement, qui est MOULINET, exécuter le premier mouvement de *à droite Moulinet.*

"Exécuter alternativement et sans s'arrêter sur aucun mouvement, le Moulinet à droite et le Moulinet à gauche."

"*À droite et à gauche* = MOULINET."

5. "*En arrière Moulinet* (1 temps, 2 mouvements). À la dernière partie du commandement, qui est MOULINET, élever le bras en arrière à droite de toute sa longueur, la pointe du sabre en l'air, le tranchant à droite, le pouce allongé sur le dos de la poignée, le corps légèrement tourné à droite."

"Décrire un circle en arrière de gauche à droite, le poignet éloigné du corps le plus possible, et se remettre en garde."

"En arrière = MOULINET."

"Les cavaliers, exécutant bien les Moulinets, on leur en fait faire plusieurs de suite, en faisant précéder cet exercice de l'indication; *les Moulinets continueront jusq'au commandement*: EN GARDE.

"Les Moulinets ayant pour objet d'assouplir les articulations du bras et du poignet, il faut que les cavaliers y soient exercés comme préparation aux autres mouvements; on commence et on finit donc chaque leçon par des Moulinets exécutés à un degré de vitesse proportionné aux progrès des cavaliers."

10 In the Regulation sword the "centre of percussion" is about one-third from the point; here there is no vibration, and consequently the Cut exercises its whole force. The "centre of gravity" is in the third nearest the hilt, and the "balance" of the sword results from the relative positions of the two centres. In light swords these points may be farther apart than in heavy blades; they should be closer in straight than in curved swords, and nearer in thrusting than in cutting weapons.

11 The following are the five principal ways of cutting:—

1. The Chopping or Downright Cut, from the shoulder and fore-arm. This appears to be the instinctive method preserved by Europe; most men who take up a sword for the first time use it in this way.

2. The Sliding Cut, common throughout the East. In this movement the elbow and wrist are held stiff and the blow is given from the strong muscles of the back and shoulder, nearly ten times larger than the muscles of the arm, while the whole force and weight of the body are thrown in. Hence the people of India use small hilts with mere crutch-guards, which confine the hand and prevent the play of the wrist; the larger grip required for the Chopping Cut only lessens the cutting force. The terrible effect of these cuts is well known.

3. The Thrust Cut, with the curved ("Damascus") blade; a combination of point and edge, the latter being obliquely thrust forward and along the body aimed at. This movement is a favourite on horseback, when speed supplies the necessary force, which can hardly be applied on foot. It must be parried like a Point.

4. The Whip Cut; in which the arm and elbow are kept almost motionless, and the blow is delivered from the wrist. This is the principal Cut allowed in my system; it is capable of sufficient effect upon the opponent whilst it does not uncover the swordsman who uses it.

5. The Drawing or Reverse Cut, which will be explained in the following pages; it is the reverse of the "Thrust Cut."

12 This fact is well known to the *Manuel*, which says, "*Des deux engagements celui de droite par la position de la main a le plus d'application.*" It therefore makes all the Cuts and Parries begin from Tierce. This elementary rule is not recognized by the 'Infantry Sword Exercise' (p. 32); "your defence is always more effective in the left (Carte) than in the right (Tierce)." Such I assert is the case with the foil and rapier, certainly not with the sabre or broadsword. On horseback the left is of course the weak side.

13 Used in this sense the "small-sword" is the triangular weapon, the rapier is the flat, or rather the bi-convex blade.

14 In p. 29 of the 'Infantry Sword Exercise' we read of "a circular motion of the blade, termed the Parry;" but the latter word must not be limited to this sense.

15 The only allusion to it is the "shifting of the leg," in p. 30.

16 See the 'Infantry Sword Exercise,' p. 31.

17 In France the false edge is hardly known; such blades are called *à deux tranchants*; it is the Italian schiena or chine, mezzo-filo, or falso opposed to vero taglio, and the German, rückschneide or kurzeschneide, thus distinguished from the lange-schneide.

18 See his Appendix, entitled "Modificazione all' impugnatura e guardia delle Sciabola di cavalleria per facilitarne l'equilibrio ed avantaggiare la fermezza della mano sull' impugnatura."

19 A notable instance of this is the old Highland Claymore.

CPSIA information can be obtained
at www.ICGtesting.com
Printed in the USA
LVHW110933221222
735718LV00005B/358

9 789356 78533